YOUR KNOWLEDGE HAS VALUE

- We will publish your bachelor's and master's thesis, essays and papers

- Your own eBook and book - sold worldwide in all relevant shops

- Earn money with each sale

Upload your text at www.GRIN.com and publish for free

Imprint:

Copyright © 2016 GRIN Verlag
Print and binding: Books on Demand GmbH, Norderstedt Germany
ISBN: 9783668820968

This book at GRIN:

https://www.grin.com/document/445685

Muhammad Haseeb

Battery Management System in an Electric Car

GRIN Verlag

BATTERY MANAGEMENT SYSTEM IN A FORMULA STUDENT ELECTRIC CAR

AUTHOR: MUHAMMAD HASEEB

DEGREE COURSE: B.ENG ELECTRICAL AND ELECTRONICS

ABSTRACT

It is believed that electric vehicles will be able to go toe-to-toe, or wheel to wheel, with internal combustion engine vehicles in as little as eight years. Some of the obstacles this industry could face is the battery lifespan and range of the vehicle. Currently, Lithium-ion batteries provide the best energy density and a longer lifespan for energy storage, however these batteries require a complex battery management system to help operate at their optimum level. This project dwells in to a basic battery management system to be used in a formula student electric car. The battery management system is custom built, making sure it adheres to the rules and regulations provided by the governing body of IMechE Formula Student. After the project completion, the system will become an integrated part of the race car which will compete at formula student UK. The possibility of it becoming the first ever British electric race car to pass scrutineering is very likely.

ACKNOWLEDGEMENTS

My project supervisor Dr. Nikola Chalashkanov of the Engineering department at the University of Leicester, UK, has been of immense help which is greatly appreciated. His generous decision to let me choose my own project has definitely made a very positive impact on my academic career as I feel this year has been the most productive one so far as I am working on something entirely of my own interest. The door to Prof. Chalashkanov's office was always open whenever I needed any guidance. He consistently allowed this paper to be my own work, but steered me in the right direction whenever he thought I needed it.

A special thanks to Mr. Rashmikanth Patel from the electronics workshop for taking out time from his schedule and helping me with the technicalities of this complex project which I couldn't have done without his help and direction. He was always there to guide me in every step to make sure this project becomes a success. I would also like to thank Dr. Paul Lefley and Luigi Alesandro for helping out the formula student team and I really appreciate their efforts in making the dream of a successful team come true.

Without their passionate participation and input, the project could not have been successfully conducted.

In the end I would like to thank my family, friends and the UOLR formula student team for all the support. Nothing could have been possible if it weren't for my parents, therefore I dedicate this thesis to my Father and Mother.

CONTENTS

Abbreviation	Definition
BMS	Battery Management System
IC	Integrated Circuit
I2C	Communication link
FET	Field effect Transistor
FSUK	Formula Student UK
AIR	Accumulator Isolation Relay
EMI	Electromagnetic Interference

BACKGROUND

Currently the batteries used in electric vehicles have a lifespan of a few hundred charge cycles after which they need replacement costing the customers a lot of money e.g. 4920£in the Nissan Leaf [10]. Another problem is the smaller range of these batteries where the Nissan Leaf only gives a range of 95 miles driving at 55 mph.

To solve these issues, an efficient battery system is required which makes sure the battery remain at optimum level prolonging its life and extending the range. Researchers at Karlsruhe Institute in Germany are testing with different battery chemistries such as using graphite nanotubes at the cathode inside the battery to increase the capacity of a battery. However, due to the same battery electrochemical properties being used over the last decade as a result of slow research processes and an increasing demand of electric vehicles, the short term solution has to come from a different approach such as a next generation battery management system.

These next generation battery management systems use highly advanced power semiconductor devices to control the critical system parameters with precision like switching power to the load instantaneously at the exact point in time. Integrated circuits with a high sensitivity and faster data links are being used in the BMS circuits to make sure all the functions are monitored and executed in real time which makes sure the battery doesn't exceed any threatening limit. Car manufacturers like Tesla Motors are using a separate thermal management system in the batteries to make sure the batteries don't need to face critical temperatures. The resulting product of all these improvements is the battery in Tesla model S with a range of 240+ miles and a longer lifecycle of its lithium batteries. [11]

All of these advancements have contributed to the success story of electric cars especially Tesla Motors which according to many experts is changing the landscape of the automobile industry. Increasing demand of electric vehicles as mentioned above was referred to what Tesla Motors has achieved with its latest Model 3 by having 200k pre-bookings of the car in just 24 hours after the prototype launch. [12]

AIMS

In an electric car, the batteries are connected to different on board systems such as the tractive and the low voltage systems. These batteries need to be more active as the car goes from acceleration to braking working in real time. Thus the battery needs to perform at the vehicle's desired operating mode which is more demanding and that is why an evolved battery management system is required. The goal was to make an easy to install, compact and stable battery management system which adheres to the rules of the FSUK competition.

In this case, the complex and costly project started from a basic grounding and a restricted budget so the only way to make it a success was to utilize what was available. The BMS boards were outsourced by the formula student team in 2013 from a PCB printing company called LEF circuits Leicester. The batteries were also purchased in the same year by the formula student team and due to high costs and budget restrictions, there was no way to purchase new batteries. Overcoming challenges with no control over the resources was a daunting task but since it had to be done, work started early on in November 2015 after completing all the paperwork required from the engineering department.

The project's essential aim is to achieve understanding of the Battery Management System and build it for an electric racecar, which will be a part of the Formula Student UK competition.Without going any further, it is important to understand the concept of a Battery Management System in an electric car which is described in the section below.

The Battery Management System in an electric car is a component of a much complex fast acting Energy Management System and must interface with other on board systems such as engine management, temperature controls, communications and safety systems. [9]

There are three main objectives common to all Battery Management Systems which are fulfilled by keeping the battery always operational in a safe operating area.

- Protect the cells or the battery from damage
- Prolong the life of the battery
- Maintain the battery in a state in which it can fulfil the functional requirements of the application for which it was specified. [9]

A BMS keeps the batteries working in the SOA (safe operating area) by:
1. Measuring Voltages & Currents*
2. Measuring temperature
3. Balancing the pack

1. Voltage & Current Measurement:
- Measures voltage and current for over/under threshold, and trips the circuit if it senses any threat to the batteries.

2. Temperature Measurement:
- Measures temperature to keep it within the safe limits during charging and discharging and trips the circuit before any damage can be occurred

3. Balancing:
- active balancing (does so by distributing charge)
- passive balancing (does so by dissipating charge through resistors) **

There are many ways of implementing a BMS. What was used in this system is the Master-Slave also called the STAR topology shown in figure (1). It consists of the boards connected in a series manner where in the functions of a master and a slave board are described below: [9]

- **The Slaves** - Each 3 cells have a temperature sensor as well as connections to measure the voltage through the sense wires, all of which are connected to the slave which monitors the condition of the cell and implements the cell balancing.
- **The Master** – 5 slaves are connected to the master board which monitors the state of the slaves. The master board controls the main battery isolation contactors (Low Voltage connection to the Accumulator Isolation Relays which provide isolation between the High voltage and Low voltage of the car) initiating battery protection in response to data from the slaves.

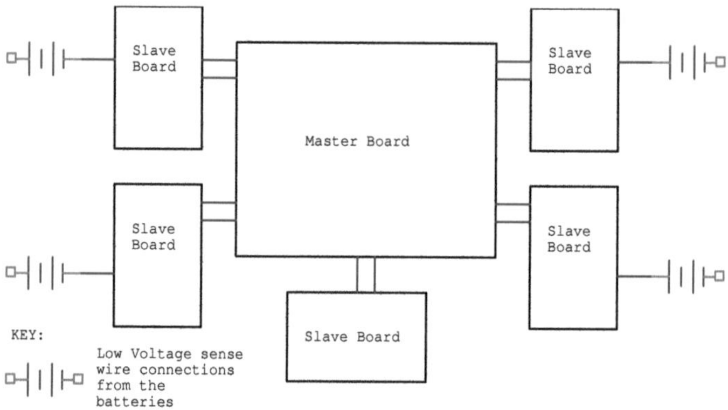

Figure 1. STAR Topology in BMS implementation.

The methodology is divided in to three categories as it would be clear to explain the function and application of each of these parts.

Battery Management System ASIC (Application specific Integrated Circuit) from Intersil, called ISL94203 (shown in figure 2) was employed on the BMS boards. The IC used is a Li-ion battery monitor IC that supports 3 to 8 series connected cells. It is a standalone (no microcontroller needed) programmable IC with Programmable detection/recovery times for overvoltage, under voltage, temperature and short circuit conditions. The boards were then populated with resistors, FETs and capacitors based on the IC configurations. For example, the charge pump capacitor pins (39,40 & 41) were connected to external capacitors and FETs used for the charge pump driving the power FETs. The BMS board can be seen in figure (3). The Intersil ISL94203 datasheet is provided so as to comprehend the functions of each register on the IC.

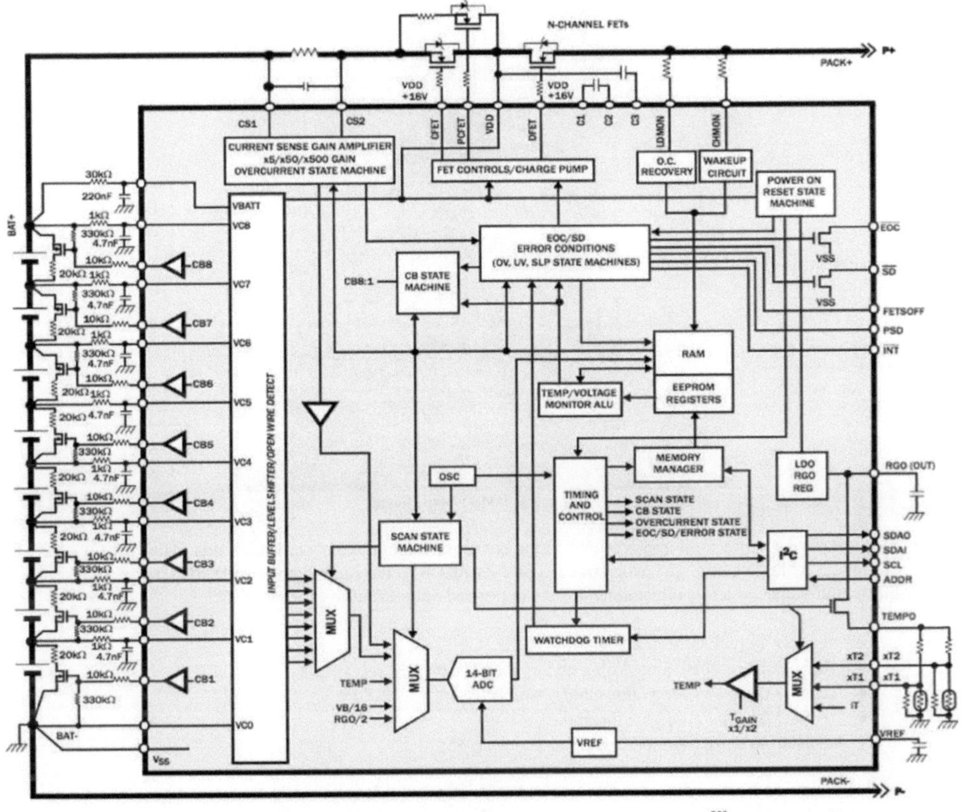

Figure 2. Battery Management System IC, ISL94203 [8]

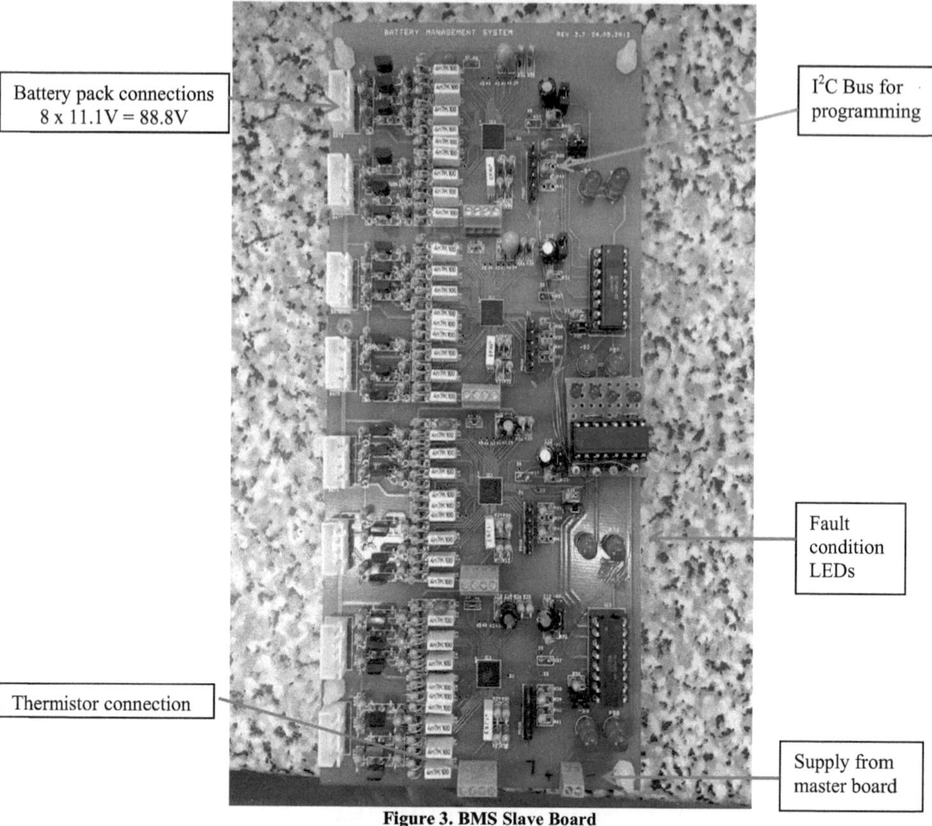

Figure 3. BMS Slave Board

A program was written and implemented on the boards through the communication link called the I2C bus on the Arduino mega. The program enables the EPROM (erasable programmable ROM) and reads the default values of a few registers which are appointed values (in the code) such as:

- register 49 (cell configuration), for 6 cells
- register 4B (cell balance during end of charge), =4.20V
- register 00 (overvoltage threshold), >4.25V
- register 04 (under voltage threshold), <3.70V
- register 30 & 38 (temperature limits), -10° to +55°

Each IC has two battery pack connections and each battery pack has 3 cells in series so monitory two battery packs meant having a configuration of 6 cells connected in series. Therefore, the first challenge was to have each IC monitor 6 cells. Register 49 handles the number of cells so a 6 cell configuration denoted by E7 (hexadecimal bit) was programmed onto it. Then the rest of the registers shown above were also programmed for the mentioned values by their corresponding hexadecimal bits.

Table 1. below show all the parameters which were programmed onto the registers. Setting the recovery threshold was important for all the parameters so that the BMS allows recovery after a fault condition has been solved. One such case is the charging process when connecting the batteries to the charger, the batteries will be drained and sit at the under voltage threshold. However, removing the load will make the battery voltage jump up hitting the recovery so that the charger LV port does not see and under voltage fault anymore. This recovery point in other words solves the under voltage fault on the BMS boards. [2]

Parameters	Value
Over voltage(&recovery) threshold	4.25V(4.08)
Cell balancing(&recovery) threshold	4.20V(4.18)
Under voltage(&recovery) threshold	3.70V(3.72)
Over temperature(&recovery) threshold	0.530V(0.590) or +55°
Under temperature(&recovery) threshold	1.344V(1.19) or -10°
Cell Configuration	6 cells connected for each IC

Table 1. Parameters set in the BMS

The next big challenge was to put the charger on standby mode while the cells were balancing. That was done by making a hardware modification shown in figure (4) to the slave board which uses an optoisolator (tlp251) to interact between the EOC and the charger by receiving a low bit on the End of Charge register from the IC and passing this signal to the charger which has a standby mode built in.

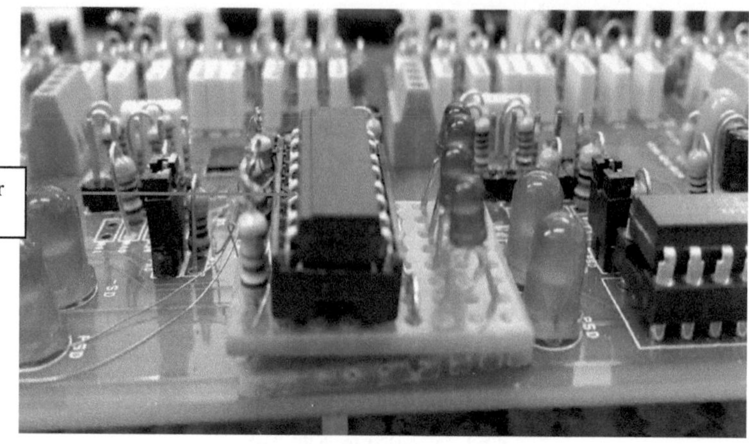

Figure 4. Charger Interaction Modification.

Another hardware modification that was required included shorting Pin 24 (serial address input to allow for multiple cascaded devices to work on a single bus) to ground, which is the serial address. The objective to short this pin with the ground is to break the cascaded link which makes every IC work independently so cell balancing occurs more efficiently between 6 cells as compared to 12 or more cells. Pins 26&27 were also shorted for the same reason which are the serial data channels acting as data lines for an I2C interface. Figure (5) shows the representation of the I2C bus with the modification.

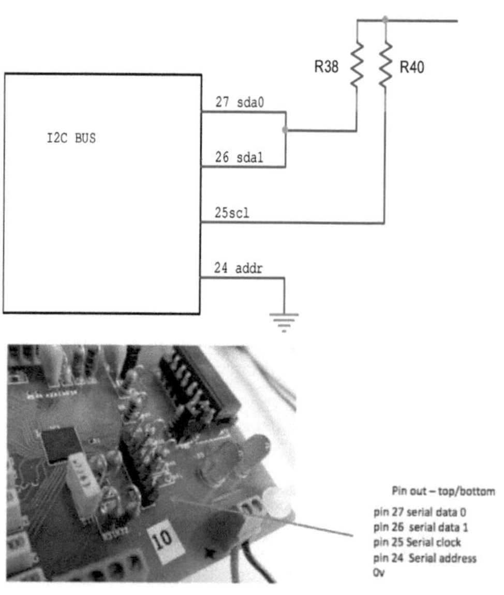

Figure 5. Communication link or I2C bus [6]

After testing and making sure the BMS slave boards were operating perfectly, a Master board shown in Figure (6) was required to cut the power from the batteries in case of a fault on the slave boards. Therefore, a comparator circuit was designed with an LM741 which is a commonly used Op-Amp. It compares 12V coming from the low voltage system of the car with a set 7.5V. The 12V would drop down because of a voltage drop across the LEDs on the master board which are turned on by the input from the slave boards. This drop gets detected by the LM741 in case of a fault and would de-energize a relay breaking the connection between the master board and the power chain (series of switches passing galvanically isolated through the HV). Thus the power from the batteries gets tripped in case the BMS slave detects any fault. LEDs inside the cockpit, alert the driver about a fault coming from the battery management system.

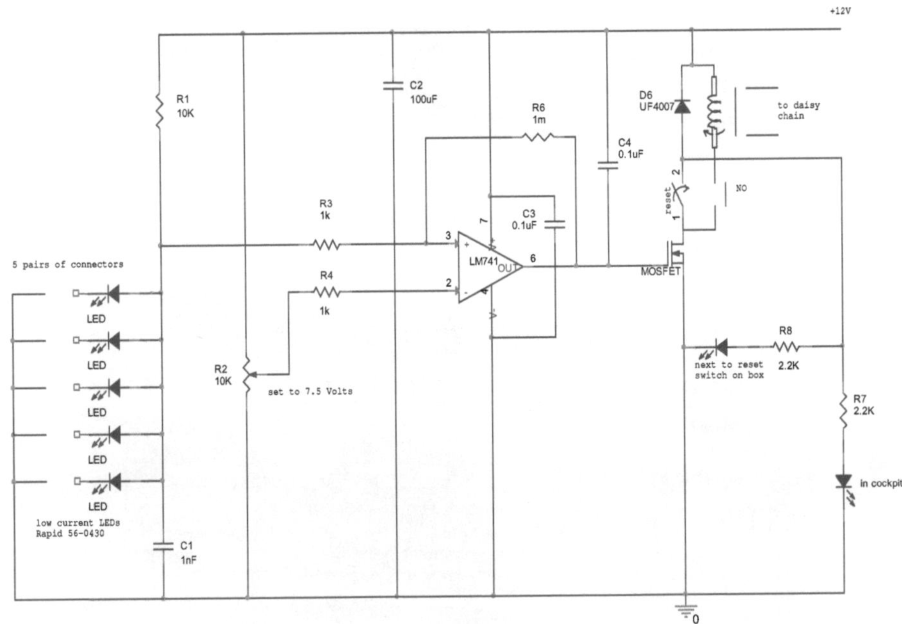

Figure 6. Master Board.

The batteries used are **Overlander 11.1V 6100mAh 3S** (3 cells in series, shown in Figure 7). The challenge was to come up with a substantive layout, which adheres to the rules and does not interrupt the safety of the battery operations. One rule states that the container should be able to resist 40g and 20g horizontal and vertical forces respectively.

Keeping these rules in mind, the container was inspired by a honey comb design where the batteries fit firm in the holders of the container resembling the honey comb blocks. Figure 8 shows the Battery container.

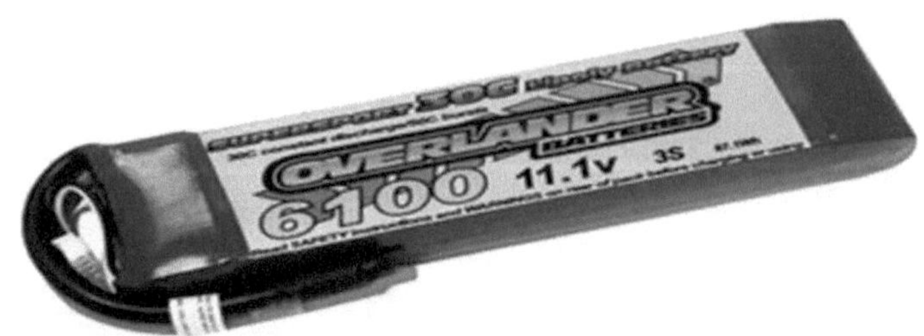

Figure 7. Overlander Supersport Series Batteries

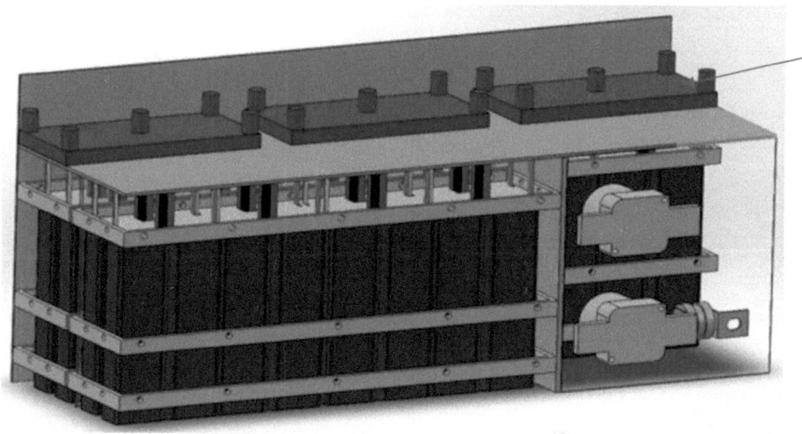

Figure 8. Accumulator/Battery Container [6]

The graphs shown below represent the battery characteristics:

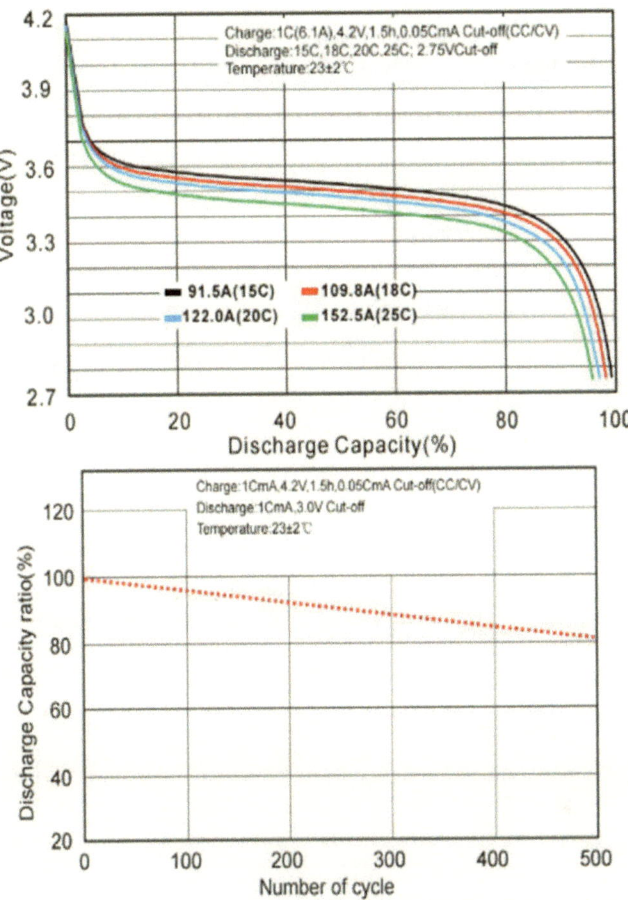

Figure 9. Battery discharge capacity and lifecycle [6]

Energy calculations done on the battery:
24 batteries in the larger segment (8 x 3)

$$6.1\,Ah\,x\,3\,=\,18.3Ah$$
$$18.3Ah\,x\,88.8V\,x\,3600s\,=\,5.85\,M \tag{1}$$

* Watt hours is energy. Multiply it by 3600 to get watt seconds which is also Joules.
Motor demand:
Power demand from the motor = 32kW

$$P\,=\,I\,x\,V \tag{2}$$
$$I\,=\,32000/88.8\,=\,360A$$

Total energy:
We have 67.6Wh batteries

$$67.7\,x\,80\,=\,5.4\,kWh \tag{3}$$

Usual range of a highway car is 3.5 miles/ kWh. Considering the power to weight ratio, it almost evens
out so rough range would be:

$$3.5\,x\,5.4\,=\,18.9\,miles$$

The electrical connection of the batteries is shown in figure 10. A total of 8 batteries are connected in series forming a string, which adds up to a voltage of 88.8V. There are 10 of these strings connected in parallel keeping the overall system voltage at 88.8V. The max charged voltage of the system can reach up to 100.8V. The fuse present in each side pod is a result of the strict rules which have to be complied. Five strings will branch out from each side pod, which will house 5 slave boards to monitor the five strings present.

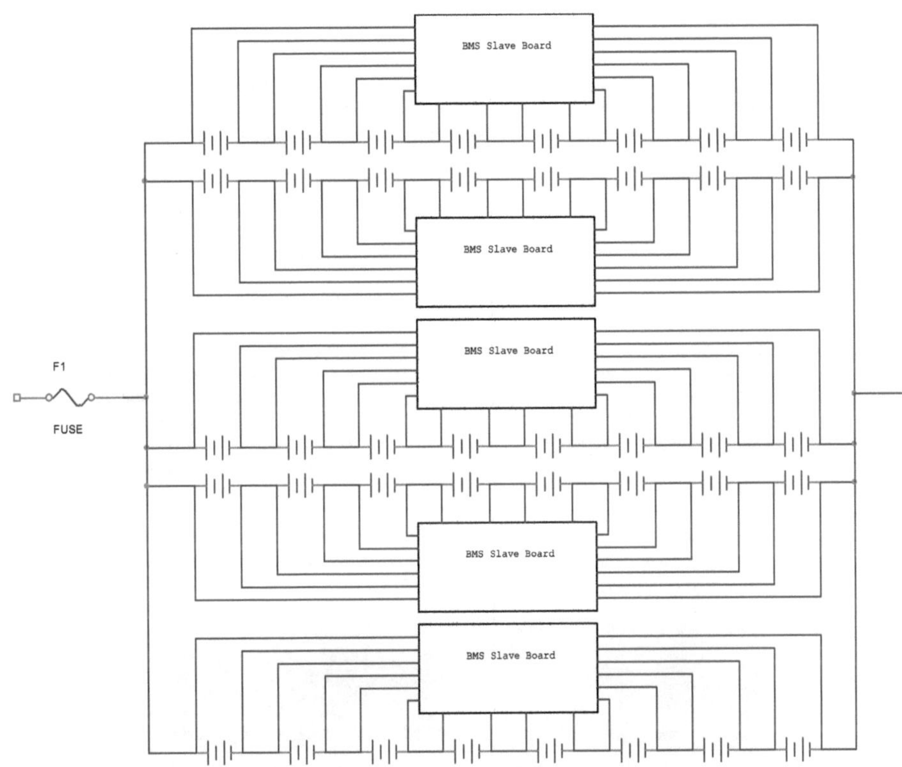

Figure 10. Electrical Representation of Batteries in each side pod

Adhering to the rules provided was a very important part of the project because the racecar needed to pass inspection based on these rules before competing in dynamic events. Below are few of the rules related to the Battery system from the FSAE website [1] and the steps taken to fulfill these requirements:

1. *EV3.6.2 The BMS must continuously measure the cell voltage of every cell, in order to keep the cells inside the allowed minimum and maximum cell voltage levels stated in the cell data sheet. If single cells are directly connected in parallel, only one voltage measurement is needed.*

   All 80 batteries (240 cells) are monitored by the 10 BMS boards, 5 in each side pod.

2. *EV3.6.3 The BMS must continuously measure the temperatures of critical points of the accumulator to keep the cells below the allowed maximum cell temperature limit stated in the cell data sheet or below 60°C, whichever is lower. Cell temperature must be measured at the negative terminal of the respective cell and the sensor used must <u>be in direct contact with either the negative terminal</u> or less than 10mm away from the terminal on the respective busbar. For lithium based cells the temperature of at least 30% of the cells must be monitored by the AMS. The monitored cells have to be equally distributed within the accumulator container.*

   Placing a thermistor at the common negative terminal of 3 cells meets the requirement of monitoring 30% of cells, in fact it monitors 33% of the cells. Figure 11 below shows the negative terminal where the thermistor is placed.

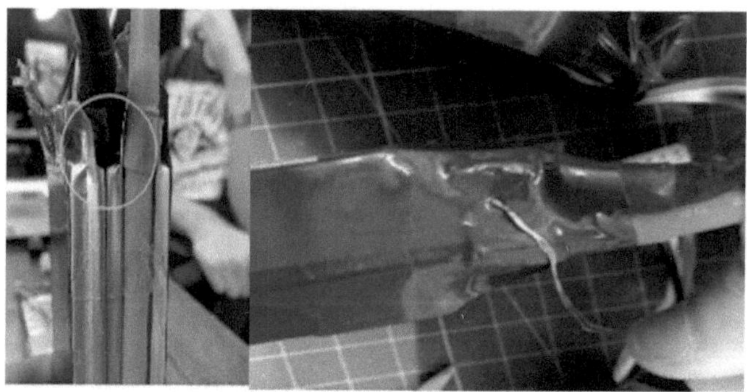

Figure 11. Battery negative terminal.

3. *EV3.6.4 For centralized BMS systems (two or more cells per BMS board), all voltage sense wires to the BMS must be protected by 'fusible link wires' or fuses so that any the sense wiring cannot exceed its current carrying capacity in the event of a short circuit.*

   320 fuses were soldered on the wires and heat shrunk as shown in **figure 12** to meet the mentioned requirement.

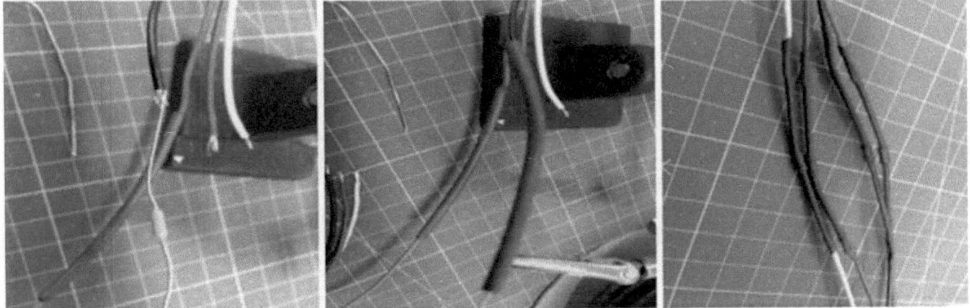

Figure 12. Sense wires fusing.

4. *EV3.6.6 NOTE: It is acceptable to monitor multiple cells with one sensor, if this sensor has direct contact to all monitored cells.*

   Each IC (sensor) on the BMS monitors 6 cells which is configured by programming through the I2C channel.

5. *EV3.6.7 The BMS must shutdown the tractive system by opening the AIRs, if critical voltage or temperature values according to the cell manufacturer's datasheet and taking into account the accuracy of the measurement system are detected. If the BMS does perform a shutdown, then a red LED marked BMS must light up in the cockpit to confirm this. NOTE: It is strongly recommended to monitor every cell temperature.*

   The circuit in **figure 6** shows the Master board connections to the GLV (grounded low voltage) system where a relay is connected which can trip in case of a fault connection lighting up and LED in the cockpit. The GLV has connections with the AIRs (Accumulator Isolation Relay) which trips the HV side.

6. *EV4.10.1 The driver must be able to (re-)activate or reset the tractive system from within the cockpit without the assistance of any other person <u>except for situations in which the BMS, IMD or BSPD have shut down the tractive system.</u>*

   This rule highlights the importance of safety when the BMS have shutdown. The driver should not be able to re-activate the BMS that's why the reset button (shown in figure 5) on the master board would be mounted on top of the container lid away from the driver.

A lot of time was spent on design that's why the manufacturing process did not face any problem and it is evident by the fact that all the sub systems work as expected. This proves the point that the design phase considered all the possibilities and shortcomings this project could face therefore actions taken were planned accordingly. One of these action included gaining knowledge of C+ with the help of staff members in the process of needing to change the coding parameters in the IC registers.

Apart from this, there were many things to learn along the way where building professional relations proved to be most lucrative. This includes negotiations with manufacturers and suppliers, learning practically with the help of staff members. Teamwork was a very important aspect in finishing this project successfully. This came in the form of collaborating with cross-functional teams like Tractive and Low Voltage system to make sure the battery system would be integrated successfully in the whole car.

Building the master board required putting designing techniques into work which were learnt through the length of this entire degree. This was also the first time where knowledge about the high voltage systems was utilized practically.

The fact that this system would be integrated into a race car which would compete at a Formula Student competition dictated a lot of aspects in this project. The most important aspect was adhering to the rules which would definitely prove beneficial if this car become the first British electric vehicle to pass electrical inspection. Due to the formal nature of the event, the governing body of the competition required the member to write protocols, qualification documents, test plans and test reports such as FMEA, ESF and design report for quality assurance purposes. This was very beneficial as these skills are considered as a very important part of a candidate's ability to succeed in the workplace.

One of the weak points in this system which could be improved includes the balancing technique. Rather than having passive balancing which releases charge through a resistor in the form of heat, active balancing seems a much better option. Active balancing is a charge displacement scheme achieved by taking charge via inductive coupling or capacitive storage from the alpha cell and injecting the stored charge into the weakest cell [8]. This way the spread of charge in the all the cells remains the same without the weakest cell going through several extra discharge cycles while dissipating heat through the resistors. The reason active balancing could not be implemented was the nature of the IC, it is a battery regulator IC without the dc-dc converters in the state machine required for active balancing.

Another imperfection this project exposed was the lack of advancement in battery management technology. One such case worth mentioning would be the inaccurate measurement of SOC (state of charge) which is an expression of the present battery capacity as a percentage of maximum capacity. SOC is generally calculated using current integration to determine the change in battery capacity over time. [3] In other words, it is always an estimate and does not tell the driver about the exact charge level or range. While this was not the most important reason of not including a display of charge level, but it did play a role considering a small range would give a high percentage error. The most important reason for omitting displaying charge level was that it requires multiple communication links on the BMS boards which take up space and are prone to EMI (electromagnetic interference).

There were many more things that were highlighted while undertaking this project. One valuable example can be the astonishing fact that cutting edge car manufacturers like Tesla Motors use the same technology, they employ BMS ICs from National Instruments and their system essentially works on a similar principle except they make it much more advanced and user friendly. This example presents a case of motivation and a lesson to learn. Former being the technology that was used in this project is industrial grade and the latter being the need for such systems to be user friendly which is the sole purpose of technology, to help humankind.

Most of the knowledge acquired in this project came thorough extensive reading and research which is the primary aim of undertaking a final year project.

The Battery Management system built will prove on to extend the shelf life of the batteries and provide safety while the car is driven. The project forms an important part of the race car which will compete at the biggest student motorsport competition in the world and this BMS will go through severe inspections by the best critiques in their field. There is a high chance of making through the inspection this year which will mean a great accomplishment not only personally, but for the university as well.

As in any professional industry, there were continually improved methods and procedures applied for processes and work flow techniques. These techniques included in-depth electrical acceptance testing of the complete battery system, including safety such as firewall and high potential isolation testing.

The results obtained are exactly what was intended which shows the amount of effort put into this project.

Technically, this system is able to perform only the basic applications of a Battery Management System because of restricted resources. However, the level it covers meets all the requirements of a fail safe system and the rules provided by the Formula Student organization. The system comprises of the most important point a battery management system is used for, which is optimizing the performance of the batteries and keeping them safe. The car shuts down as soon as the batteries start to posses a threat in terms of exceeding voltage limits, temperature limits or causing a short circuit condition. The fuses incorporated between the batteries take care of the current limits which is part of the FSUK rules and also seemed as an easier alternative to having current monitoring in the BMS.

In spite of being a basic level project, advancement in such work could evolve the future of energy storage which is fast changing. Better batteries and more advanced systems of managing them are required at the point where the auto industry is on the verge of switching polarity going from gasoline to electrical. Some start up businesses like Faradion UK are already testing with different battery chemistries like sodium ions instead of lithium ions combined with next generation energy management systems. This will reduce the battery costs and give a big push to the electric car industry. This is much bigger than the automobile industry, it would have a huge impact on the planet as a whole.

REFERENCES

[1] Auto express, 'Nissan Leaf battery replacement to cost £4,920', 8th Dec 2014.
(http://www.autoexpress.co.uk/nissan/89694/nissan-leaf-battery-replacement-to-cost-4920)

[2] David Andrea, 'Battery Management Systems for Large Lithium-Ion Battery Packs', 1st Edition, 2010.

[3] EV Obsession, 'Electric Vehicles with most range', 29th July 2015.
(http://evobsession.com/electric-car-range-comparison/)

[4] Electropaedia, 'Battery and Energy Technologies'.
(http://www.mpoweruk.com/bms.htm)

[5] Forbes, 'Tesla Model 3 Orders Wow Advocates; Skeptics, Not So Much', 7th April 2016.
(http://www.forbes.com/sites/michaellynch/2016/04/07/tesla-model-3-orders-wow-advocates-skeptics-not-so-much/#39431dd87ffd)

[6] Formula SAE® Rules, 2015
(http://www.fsaeonline.com/content/2015-16%20FSAE%20Rules%20revision%2091714%20kz.pdf)

[7] Google Drive, University of Leicester Racing Team

[8] Intersil isl94203 datasheet.
(http://www.intersil.com/content/dam/Intersil/documents/isl9/isl94203.pdf)

[9] JT Baer, BC Davis, RJ Blanyer- Google Patents, US Patent 5,701,068, 1997.

[10] MIT Electric Vehicle Team, 'A Guide to Understanding Battery Specifications', December 2008.

[11] Phillip Weicker, 'A Systems approach to Lithium-ion Battery Management', 1st Edition, 2013.

[12] Ralph E. Scheidler, 'Introduction to batteries and charging systems', Revised Edition, 1994.

EXPERIMENT 1

All the BMS boards were programmed and tested for working in a carefully designed experiment described briefly below:
- connect the battery simulator to the BMS board.
- Make sure the thermistor connections and the sense wire connections are correct on the BMS board.
- connect the Arduino Mega to the laptop and the I2C bus on the BMS board.
- implement the codes on to the boards through the I2C bus.
- Check voltage thresholds on Arduino screen print by varying ports on the battery simulator.
- Any fault detected should switch the LEDs on.

BATTERY PACK SPECIFICATIONS

- Voltage 11.1V
- Current rating 6100mAh- Amp hours is a measure of the capacity of a cell. In this case, this cell could provide 1A for 2.2 hours at a continuous voltage of 11.1V
- 67.1 WH capacity- Watt hour is a measure of power expended every one hour
- Dimensions: Length 180 x Width 50 x Height 30mm
- Weight 0.460 kg

CAR DESIGN [6]

Figure 12. Side View of the Car showing one of the battery Side pod